HARCOURT Math
GEORGIA EDITION

Challenge Workbook

Grade K

Harcourt

Visit The Learning Site!
www.harcourtschool.com

Copyright © by Harcourt, Inc.

All rights reserved. No part of this publication may be reproduced or transmitted in any form or by any means, electronic or mechanical, including photocopy, recording, or any information storage and retrieval system, without permission in writing from the publisher.

Permission is hereby granted to individual teachers using the corresponding student's textbook or kit as the major vehicle for regular classroom instruction to photocopy complete pages from this publication in classroom quantities for instructional use and not for resale. Requests for information on other matters regarding duplication of this work should be addressed to School Permissions and Copyrights, Harcourt, Inc., 6277 Sea Harbor Drive, Orlando, Florida 32887-6777. Fax: 407-345-2418.

HARCOURT and the Harcourt Logo are trademarks of Harcourt, Inc., registered in the United States of America and/or other jurisdictions.

Printed in the United States of America

ISBN 13: 978-0-15-349550-2
ISBN 10: 0-15-349550-2

If you have received these materials as examination copies free of charge, Harcourt School Publishers retains title to the materials and they may not be resold. Resale of examination copies is strictly prohibited and is illegal.

Possession of this publication in print format does not entitle users to convert this publication, or any portion of it, into electronic format.

2 3 4 5 6 7 8 9 10 054 15 14 13 12 11 10 09 08 07

CONTENTS

Chapter 1: Sort and Classify
1.1 On the Shelves 1
1.2 Where is It? 2
1.3 The Picnic 3
1.4 A Day at the Park 4
1.5 Colors and Shapes 5
1.6 What Kind Is It? 6
1.7 Letter Sorting 7
1.8 Pets 8

Chapter 2: Patterns
2.1 Patterns of Sounds 9
2.2 Patterns to Wear 10
2.3 What Shape Comes Next? 11
2.4 Autumn Patterns 12
2.5 Pattern Notes 13
2.6 Pup Patterns 14
2.7 Circle and Square Patterns 15
2.8 Draw the Missing Blocks 16

Chapter 3: Numbers 0 to 5
3.1 More to Match 17
3.2 More Toys 18
3.3 Fewer Animals 19
3.4 Nature Graphing 20
3.5 Shapes 21
3.6 Amazing 5 22
3.7 How Many Balls? 23
3.8 Gum Ball Estimation 24

Chapter 4: Numbers 6 to 10
4.1 Groups of 6 and 7 25
4.2 Shape Up 26
4.3 Buzz on to 10 27
4.4 Cube Trains 28
4.5 Share Objects Equally 29
4.6 What is My Number? 30
4.7 Number Combinations 31
4.8 Guessing Jar 32

Chapter 5: Geometry
5.1 Sort it Out 33
5.2 What Shape Am I? 34
5.3 See the Shape 35
5.4 Shapes to Color 36
5.5 Different Shapes,
     Different Places 37
5.6 Combine Plane Shapes 38
5.7 Represent Plane Shapes 39

Chapter 6: Numbers 10 to 30
6.1 Animals First through Fifth 40
6.2 Ten Frame Color 41
6.3 Fill a Frame 42
6.4 Ten Frame Teens 43
6.5 Ten and More 44
6.6 Old MacDonald's Farm 45
6.7 Two for Twenty 46
6.8 Missing Numbers 47

Chapter 7: Money
- 7.1 How Many Pennies? 48
- 7.2 Trading 49
- 7.3 Nickel or Dime 50
- 7.4 Quarter 51
- 7.5 Toy Sale 52
- 7.6 Problem Solving Strategy: Use Objects 53

Chapter 8: Time
- 8.1 What is Missing? 54
- 8.2 Put Them in Their Place 55
- 8.3 What Time Is It? 56
- 8.4 Busy Week 57
- 8.5 Missing Months 58
- 8.6 Colors on a Calendar 59
- 8.7 Outdoor Fun 60

Chapter 9: Measurement
- 9.1 Measurement in the Classroom 61
- 9.2 I Can Draw It 62
- 9.3 Up, Up, and Away 63
- 9.4 Which Object is Tallest? 64
- 9.5 Capacity in the Refrigerator 65
- 9.6 Which Object is Lightest? 66

Chapter 10: Data and Graphing
- 10.1 Shape Graph 67
- 10.2 Fruit Graph 68
- 10.3 Let's Play 69
- 10.4 Clothes Count 70
- 10.5 Fruit Stand 71
- 10.6 Picnic in the Park 72

Chapter 11: Addition
- 11.1 Ways to Make 5 73
- 11.2 Making 6 and 7 74
- 11.3 Making 10 75
- 11.4 Making 8 76
- 11.5 Circle How Many in All 77
- 11.6 Add, Add, Add 78
- 11.7 Adding with Pictures 79
- 11.8 Trade a Story 80

Chapter 12: Subtraction
- 12.1 Take It Away! 81
- 12.2 How Many are Left? 82
- 12.3 Making Subtraction Patterns 83
- 12.4 Dot Subtraction 84
- 12.5 Subtraction Sentences 85
- 12.6 Drawing Pennies 86
- 12.7 Supermarket Subtraction 87

© Harcourt

Name _____

● On the Shelves

Draw and color a ball above the bear on the middle shelf. Draw and color a ball beside the bear on the middle shelf. Draw and color a hat below the bear on the middle shelf.

Challenge   CW1

Name _____

▶ **LESSON 1.2**

Where is It?

Inside   Outside   Inside   Outside

Inside   Outside   Inside   Outside

Circle the correct position word.
- 🐟 Is the soccer ball inside or outside of the goal?
- 🐢 Is the dog inside or outside of the dog house?
- ★ Is the flower inside or outside of the vase?
- ♥ Are the grapes inside or outside of the bowl?

CW2 Challenge

Name _____

LESSON 1.3

● The Picnic

Use yellow to color the object in front of the tree. Use brown to color the object behind the tree. Draw one more object in front of the tree.
Draw a girl behind the boy playing volleyball.
Use purple to color the object in front of the table. Draw a boy behind the picnic table.

Challenge   CW3

Name _____

LESSON 1.4

A Day at the Park

Find the tree that is in front of the fence. Use green to color the top, brown to color the middle, and orange to color the bottom.
Find the sign. Draw and color a red bird above the sign. Draw and color a blue bird below the sign.
Use purple to draw and color a bench beside the path inside the park.

CW4   Challenge

Name _____

LESSON 1.5

Colors and Shapes

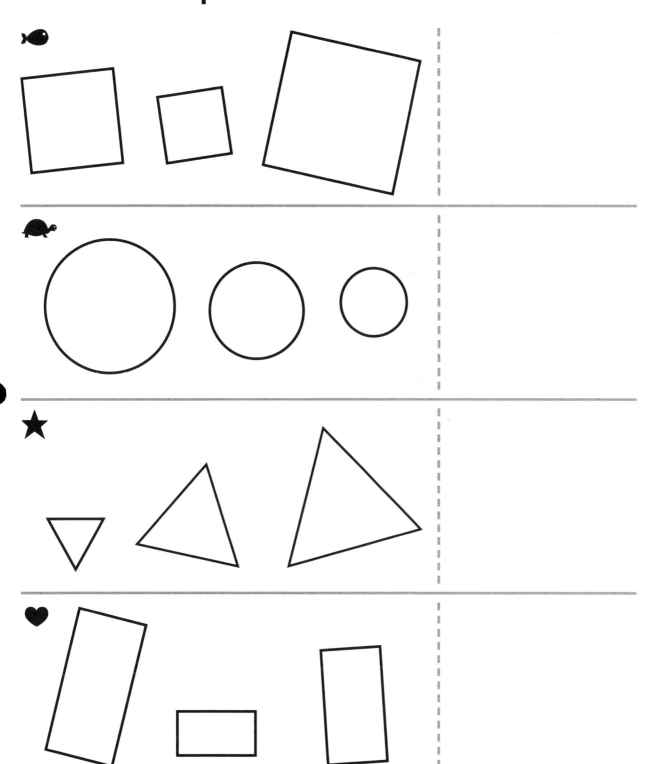

🐟 🐢 ★ ♥ Choose a color. Color each shape in the group. Draw and color one more shape to match.

Challenge CW5

Name _____

What Kind Is It?

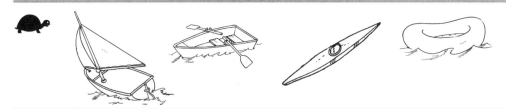

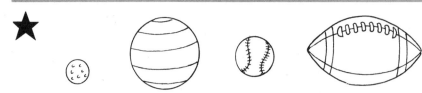

- 🐟 Use yellow to color the object in the toy box that is the same kind.
- 🐢 Use red to color the object in the toy box that is the same kind.
- ★ Use blue to color the object in the toy box that is the same kind.

CW6   Challenge

Name _____

LESSON 1.7

● **Letter Sorting**

L E O F

V W M C

X C S O

H S E K

● 🐟 🐢 ★ ♥ Sort the letters so 3 of them are alike. Which letter does not belong? Mark an **X** on it and tell why.

Challenge   CW7

Name _____

▶ LESSON 1.8

Pets

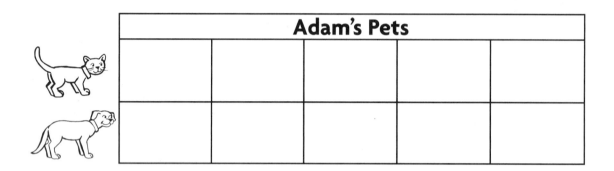

 Look at the picture. Color one box for each pet. Does Adam have more cats or more dogs? How do you know?

 Look at the picture. Color one box for each pet. Does Amy have more fish or more hamsters? How do you know?

CW8 Challenge

Patterns of Sounds

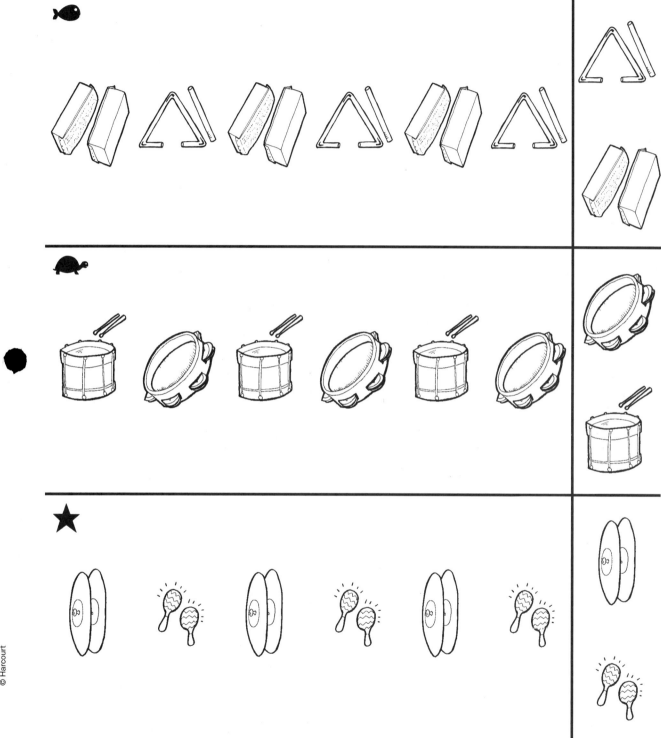

Think about the pattern of sounds. Circle the instrument that you would most likely hear next.

Challenge CW9

Name _____

 LESSON 2.2

Patterns to Wear

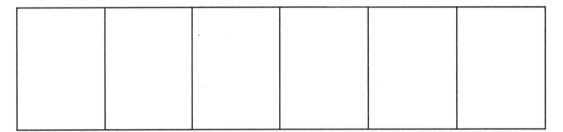

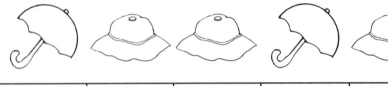

 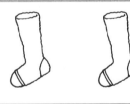 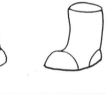

🐟 🐢 ★ Copy the pattern in the empty boxes.

CW10 Challenge

Name _____

LESSON 2.3

What Shape Comes Next?

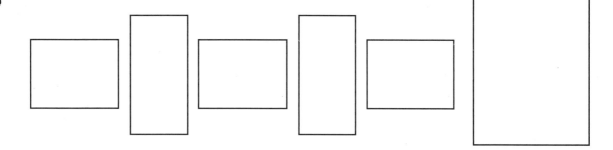

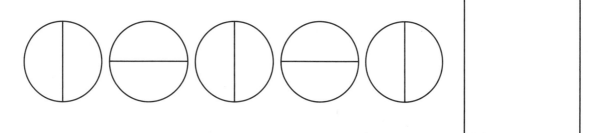

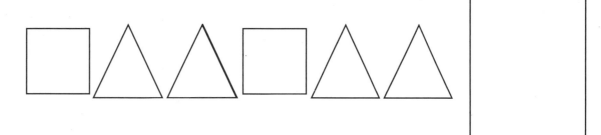

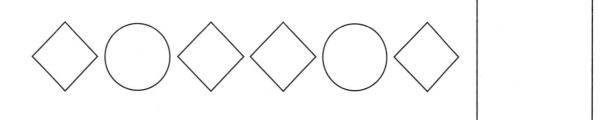

🐟 🐢 ★ ♥ Draw the shape that most likely comes next.

Challenge CW11

Name _____

Autumn Patterns

Predict, then draw to extend what most likely comes next in the pattern.

CW12 Challenge

Name _____

● Pattern Notes

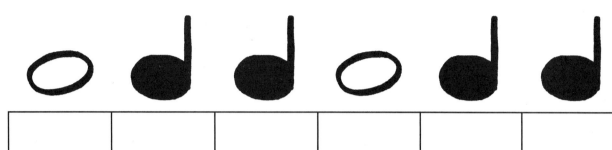

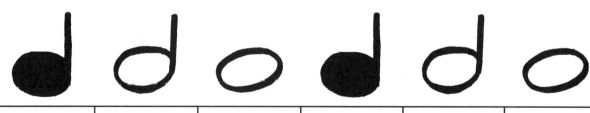

🐟 🐢 ★ Use classroom instruments to copy the pattern. Draw the instruments you use.

Challenge CW13

Name _____

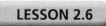 LESSON 2.6

Pup Patterns

Circle the part of the pattern that repeats again and again.

CW14 Challenge

Name _____

LESSON 2.7

● Circle and Square Patterns

● 🐟 🐢 ★ ♥ Use circles and squares to draw your own pattern in each row.
Then share your patterns with a friend.

Challenge CW15

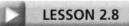

Draw the Missing Blocks

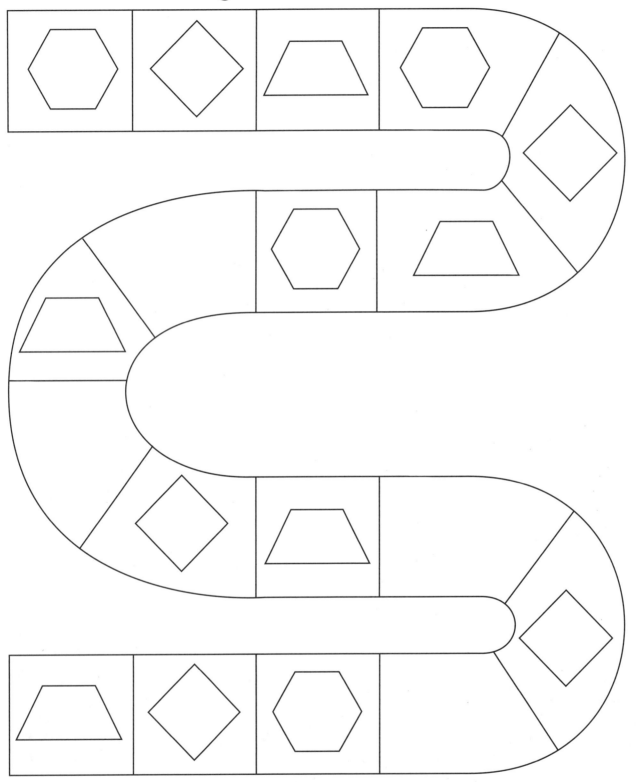

Read the pattern. Tell what shapes repeat again and again. Draw the missing shapes in the pattern. Color each of the three shapes a different color.

CW16 Challenge

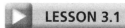
LESSON 3.1

● More to Match

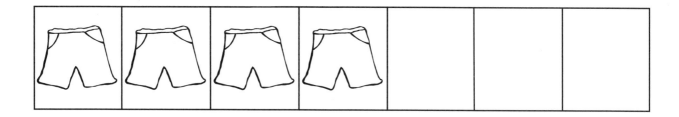

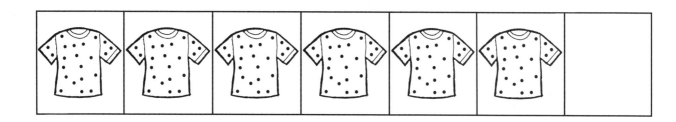

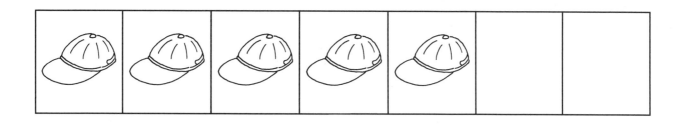

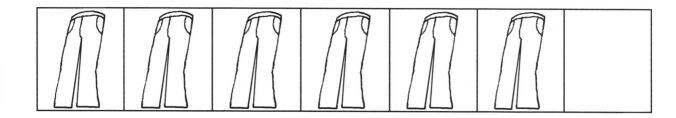

● Circle the two rows that have the same number.
Draw more objects to make all the rows have that number.

Challenge   CW17

Name _____

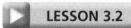 **LESSON 3.2**

More Toys

Draw a set that has more toys.

CW18 Challenge

Name _____

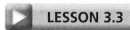

● Fewer Animals

● 🐟 🐢 ★ ♥ Draw a set that has fewer animals.

Challenge   CW19

Name _____

LESSON 3.4

Nature Graphing

Birds and Leaves

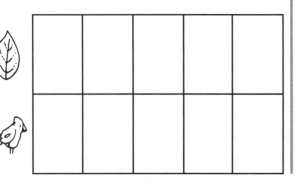

Flowers and Bees

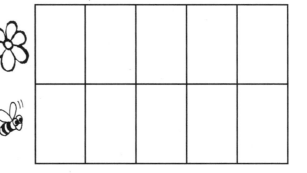

- Draw the objects on the graph. Circle the row that shows more.
- Draw the objects on the graph. Circle the row that shows more.

CW20   Challenge

Name _____

● Shapes

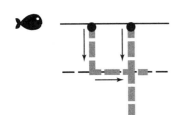 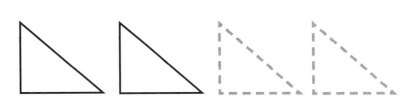

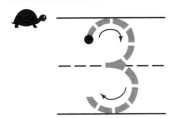 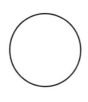

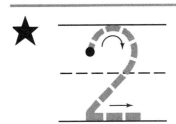

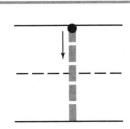

- 🐟 Trace the number. Draw more triangles to model the number.
- 🐢 Trace the number. Draw more circles to model the number.
- ★ Trace the number. Draw triangles to model the number.
- ♥ Trace the number. Draw a circle to model the number.

Challenge CW21

Name _____

LESSON 3.6

Amazing 5

 3 4 5

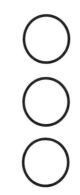 3 4 5

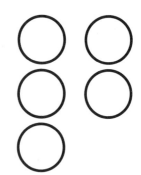 3 4 5

  3 4 5

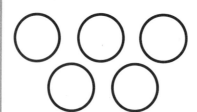 3 4 5

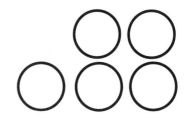 3 4 5

★ 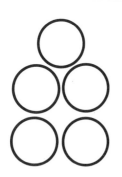 3 4 5

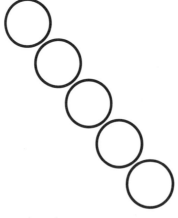 3 4 5

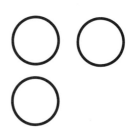 3 4 5

🐟 🐢 ★ Circle the number that tells how many. Color the groups of five.

CW22 Challenge

Name _____

LESSON 3.7

How Many Balls?

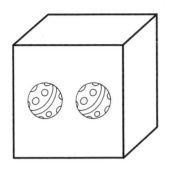
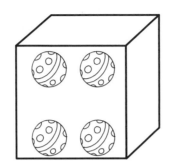
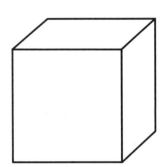

_____   _____   _____

-------   -------   -------

_____   _____   _____

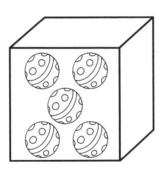
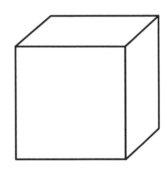
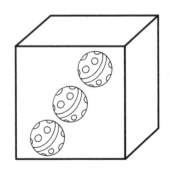

_____   _____   _____

-------   -------   -------

_____   _____   _____

Count the balls and write how many.

Challenge   CW23

Gum Ball Estimation

less than 10

10

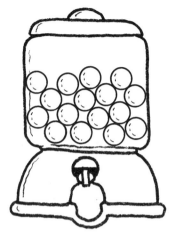
more than 10

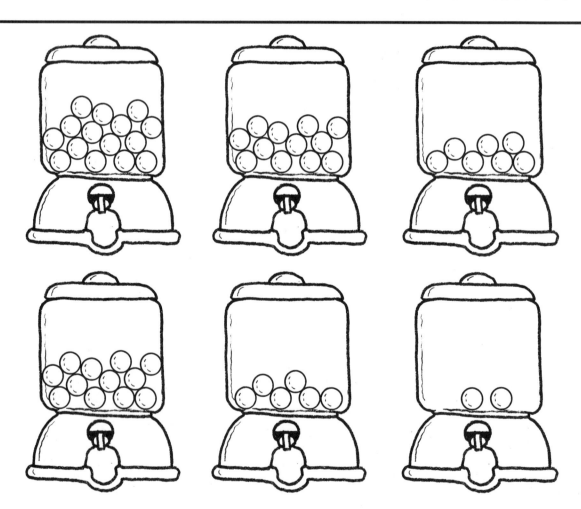

Circle the pictures that show more than 10.
Mark an X on the pictures that show less than 10.

Name _____

LESSON 4.1

● Groups of 6 and 7

 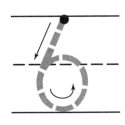 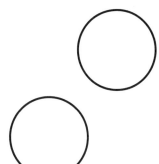

 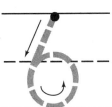

●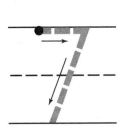

 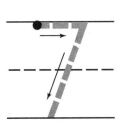 

● 🐟 🐢 Write the number. Draw more objects to make a group of 6.
★ ♥ Write the number. Draw more objects to make a group of 7.

Challenge CW25

Name _____

LESSON 4.2

Shape Up

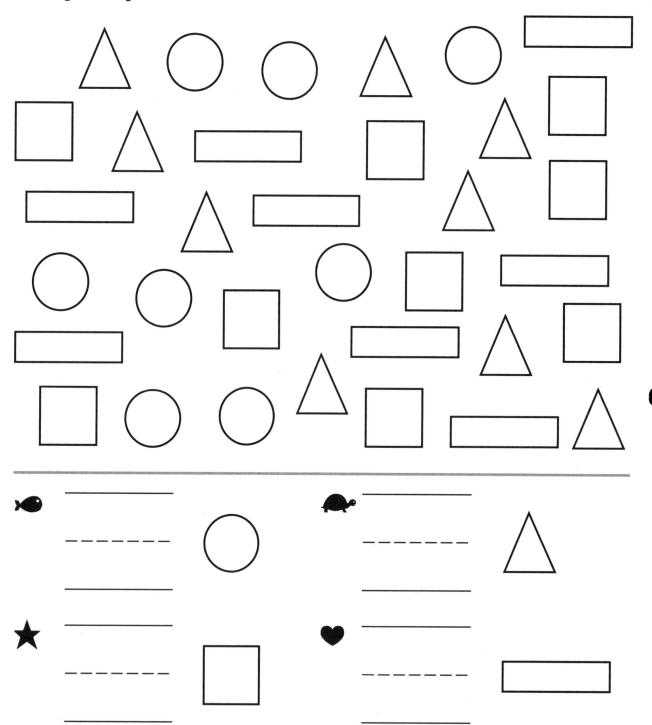

- 🐟 Count the circles in the picture. Write the number.
- 🐢 Count the triangles in the picture. Write the number.
- ★ Count the squares in the picture. Write the number.
- ♥ Count the rectangles in the picture. Write the number.

CW26 Challenge

Name _____

LESSON 4.3

Buzz on to 10

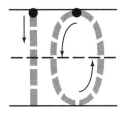

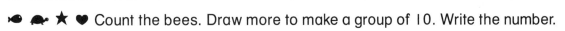 Count the bees. Draw more to make a group of 10. Write the number.

Challenge CW27

Cube Trains

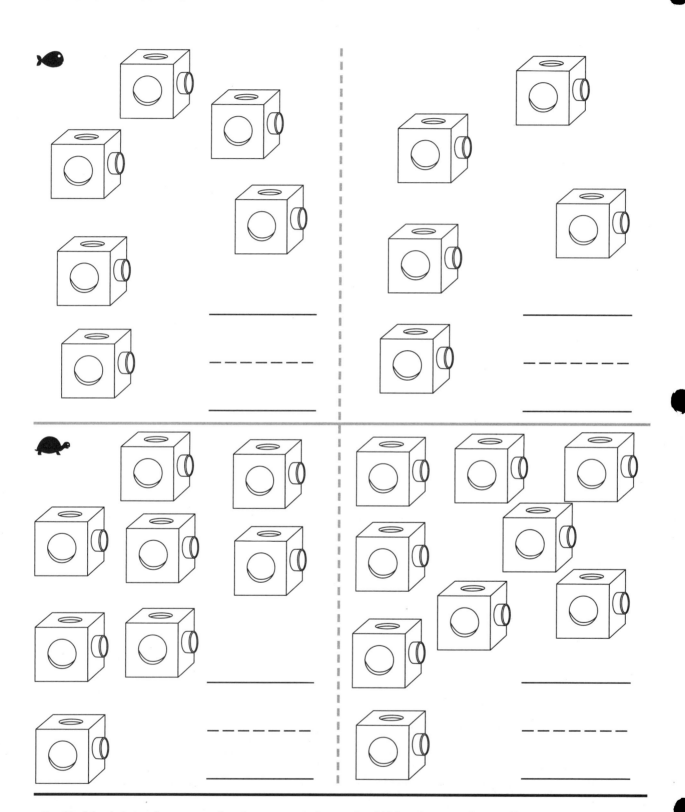

🐟 🐢 Model each group of cubes as a cube train. Write the numbers. Compare the cube trains. Circle the number that shows more. How did you know?

CW28 Challenge

Share Objects Equally

 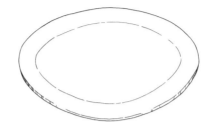 

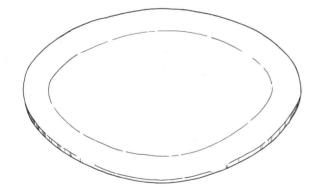 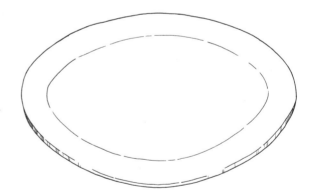

 Which group of cookies can be shared equally on the plates? Circle the group of cookies and draw the same number on each plate.

Challenge   CW29

Name _____

What is my Number?

🐟 🐢 ★ ♥ Look at the numbers on the shirts. Write the missing numbers.

Name _____

LESSON 4.7

Number Combinations

blue _____ _____ yellow

blue _____ _____ yellow

blue _____ _____ yellow

Use blue and yellow crayons to color 3 different ways to make ten. Write how many of each color.

Challenge CW31

Guessing Jar

The jar at the top of the page has 4 marbles. Without counting, find groups of marbles that have more than 4. Circle the groups with more than 4 marbles.

CW32   Challenge

Name _____

  LESSON 5.1

Sort It Out

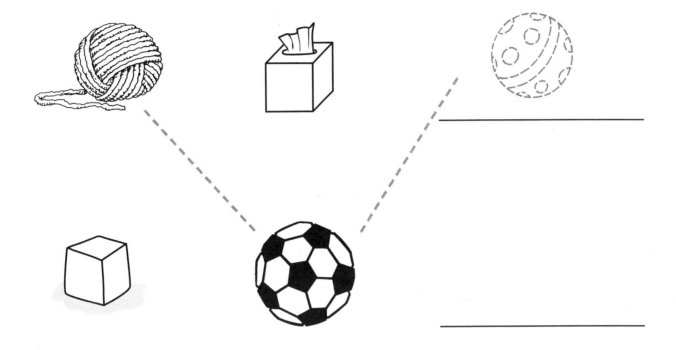

 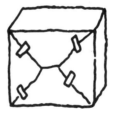

Look at the solid figures. Draw one more of each shape. Draw lines to connect the shapes that are alike.

Challenge CW33

What Shape Am I?

sphere

cube

🐟 I have six flat surfaces.
Every surface is flat with equal sides.
I slide and cannot roll.
What shape am I?

🐢 I am curved all over.
I can roll.
What shape am I?

🐟 🐢 Listen to the riddle. Answer the question and draw the shape.

CW34 Challenge

See the Shape

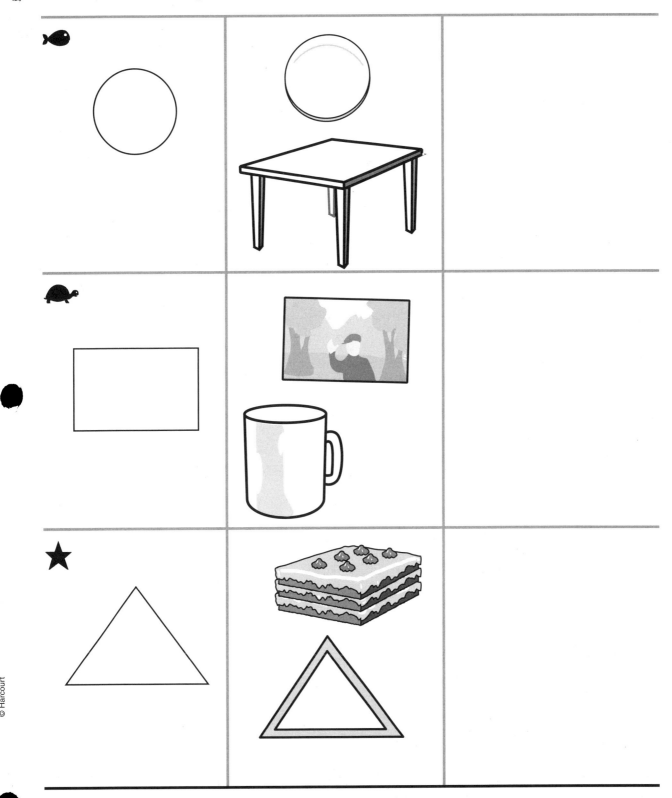

🐟 🐢 ★ Look at the outline. Circle the object that matches the outline. Then draw another outline that matches the object.

Challenge   CW35

Name _____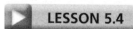

Shapes to Color

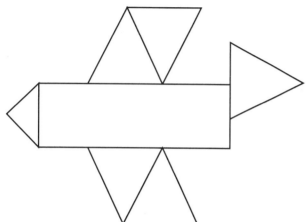

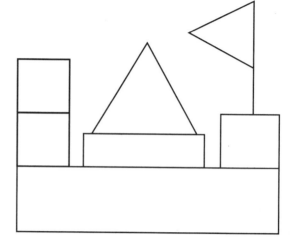

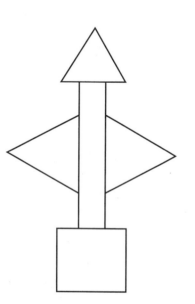

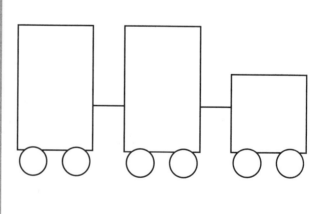

 Color each ▭ blue.
Color each △ yellow.
Color each ○ red.
Color each ▢ green.

CW36 Challenge

Name _____

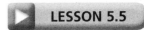

Different Shapes, Different Places

Color the boxes to show how many circles, squares, rectangles, and triangles.

Challenge   CW37

Name _____

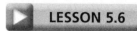

Combine Plane Shapes

   Use 3 or more plane shapes to create a new figure. Color the figure.

CW38 Challenge

Name _____

LESSON 5.7

● **Represent Plane Shapes**

●

★

● Use the plane shapes to make a picture of a train.
    Use the plane shapes to make a picture of a person.
   ★ Use the plane shapes to make a picture of a tower.

Challenge   CW39

Name _____

LESSON 6.1

Animals on Parade

Use red to color the first animal. Use yellow to color the third animal. Use blue to color the sixth animal. Use green to color the tenth animal.

CW40 Challenge

Name _____

● Ten Frame Color

 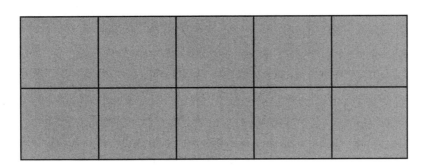 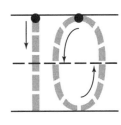

● ★

● ♥

 Color in the ten frame to show the number that is one more than nine.
Write the number.

 Color in the ten frame to show the number that is one less than ten. Write the number.

★ Color in the ten frame to show the number that is three more than seven.
Write the number.

♥ Color in the ten frame to show the number that is two less than ten. Write the number.

Challenge CW41

Name _____

  LESSON 6.3

Fill a Frame

 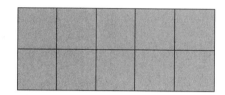 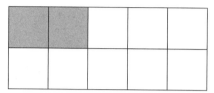

 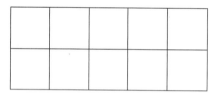 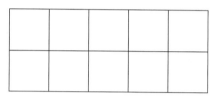

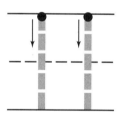 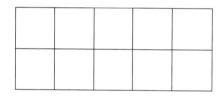 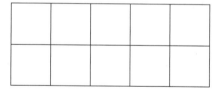

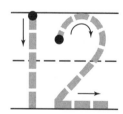 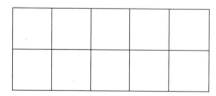

🐟 🐢 ★ ♥   Trace the number. Fill in the ten frames to show that number.

CW42   Challenge

Name _____

Ten Frame Teens

🐟
14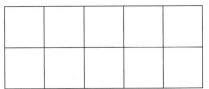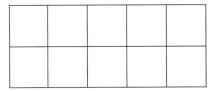

🐢
15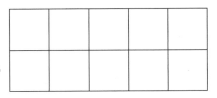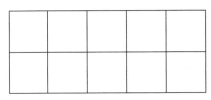

★
16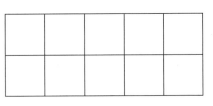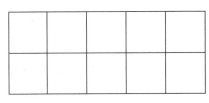

♥
15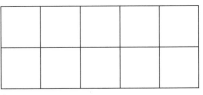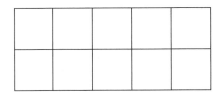

🐟 🐢 ★ ♥   Read the number. Draw the object in the ten frames to show that number.

Challenge CW43

Name _____ 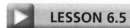 LESSON 6.5

Ten and More

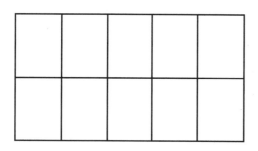 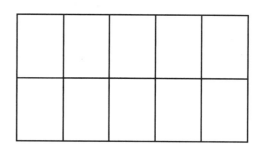

ten and ____ more

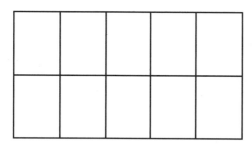 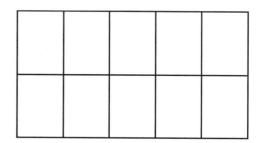

ten and ____ more

Read each number. Fill in the ten frames to show ten and how many more.

CW44 Challenge

Name _____

Old MacDonald's Farm

Animals On the Farm

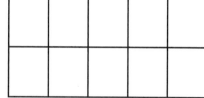

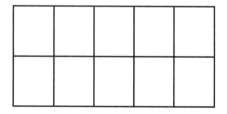　

　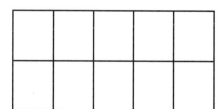

Count the number of each animal on the picture graph. Write how many. Fill in the ten frames to show that number.

Challenge　CW45

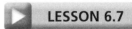

Name _____

Two for Twenty

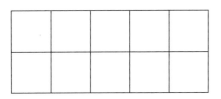

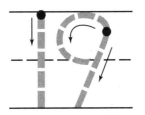

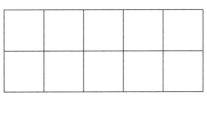

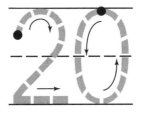

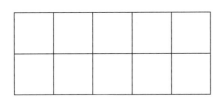

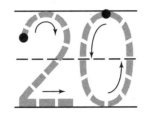

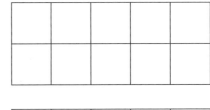
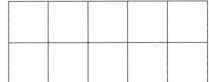
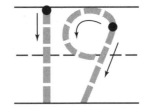

🐟 🐢 ★ ♥ Trace the number. Fill in the ten frames to show each number.

CW46 Challenge

Name _____

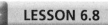 **LESSON 6.8**

Missing Numbers

←― 21 22 23 24 25 26 27 28 29 30 ―→

🐟
| 23 | | 25 |

22   26   (24)

🐢
| 26 | | 28 |

25   27   29

★
| 24 | 25 | |

24   26   28

♥
| 28 | | 30 |

21   23   29

✿
| 22 | | 24 |

23   25   26

🦋
| 28 | 29 | |

20   25   30

🐟 🐢 ★ ♥ ✿ 🦋 Circle the missing number.

Challenge CW47

How Many Pennies?

13 ¢

____ ¢

____ ¢

____ ¢

🐟 🐢 Count the pennies. Write the value.
★ ♥ Mark an X on the coins that are not pennies. Count the pennies. Write the value.

CW48   Challenge

Trading

Draw nickels to show the same value as the group of pennies.

Name _____

LESSON 7.3

Nickel or Dime?

Circle the coin that is equal to the value of the group of pennies.

CW50   Challenge

Name _____

LESSON 7.4

Quarter Pick

🐟
25¢

🐢
10¢

★
5¢

♥
25¢

🐟 🐢 ★ ♥ Circle the coin that equals the amount shown.

Challenge CW51

Name _____

LESSON 7.5

Toy Sale

🐟

_ _ _ _ _ _ ¢

🐢

_ _ _ _ _ _ ¢

🐟 Choose a toy to buy. Circle the toy. Draw the coins you would use to buy the toy. Write the amount.
🐢 Draw different coins you could use to buy the same toy. Write the amount.

CW52 Challenge

Name _____

LESSON 7.6

Problem Solving Strategy: Use Objects

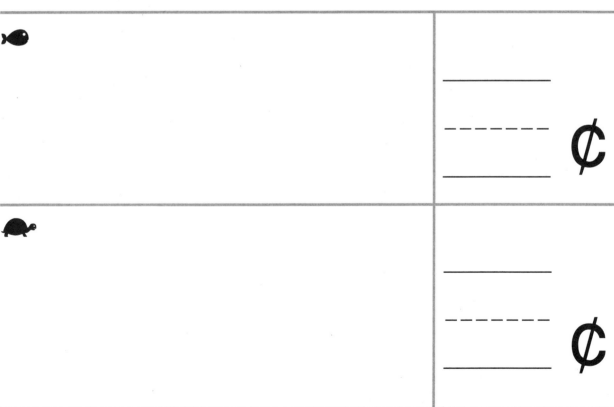

🐟 Choose a toy to buy. Show the coins you would use to buy the toy. Write how many cents.
🐢 Trade the coins and show another way to buy the toy. Write how many cents.

Challenge   CW53

Name _____

LESSON 8.1

What is Missing?

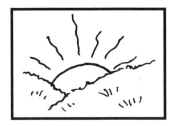

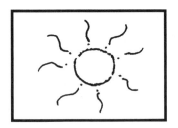

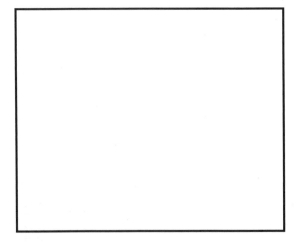

Draw a line to match each picture to its time of day. Circle the time of day that has no picture. Then draw a picture in the box for that time of day.

CW54 Challenge

LESSON 8.2

Name _____

Put Them in Their Place

🐟 🐢 ★ Draw a picture to show what might happen first, second, or third.

Challenge CW55

Name _____

What Time Is It?

   Circle the time when each event happens.

Name _____

LESSON 8.4

Busy Week

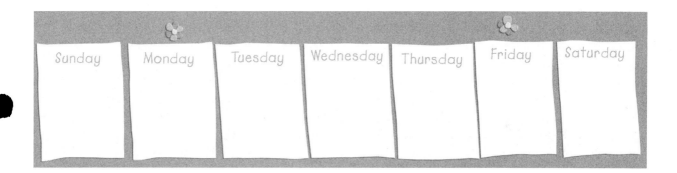

Tom plays baseball every Tuesday. Draw a baseball bat in the box for Tuesday.

He has swim lessons on Thursday. Draw a pool in the box for Thursday.

He plays outside on Monday and Wednesday. Draw a sun in the boxes for Monday and Wednesday.

Tom's family has a party on Saturday. Draw a cake in the box for Saturday.

Challenge   CW57

Name _____

LESSON 8.5

Missing Months

Use yellow to color the first month of the year.

Use green to color the month with the least number of days.

Use red to color the month it is now.

Use blue to color the month that you begin school each year.

CW58 Challenge

Name _____

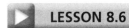 LESSON 8.6

Colors on a Calendar

March

Sunday	Monday	Tuesday	Wednesday	Thursday	Friday	Saturday
	1	2	3	4	5	6
			10			
	15					20
				25		
		30	31			

Write the missing numbers on the calendar.
Use red to color the numbers that are between 2 and 6.
Use blue to color the numbers that are between 14 and 22.

Challenge   CW59

Name _____

LESSON 8.7

Outdoor Fun

Use blue to color the best activity for winter.

Use green to color the best activity for spring.

Use orange to color the best activity for fall.

Use yellow to color the best activity for summer.

CW60 Challenge

Name _____

LESSON 9.1

Measurement in the Classroom

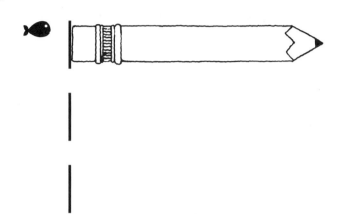

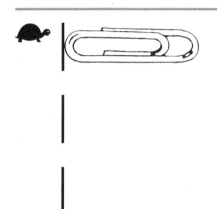

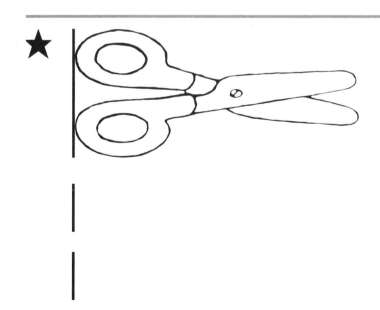

🐟 🐢 ★ Draw an object that is longer than the one you see. Next draw an object that is shorter than the one you see. Then circle the longest object.

Challenge   CW61

Name _____

LESSON 9.2

I Can Draw It

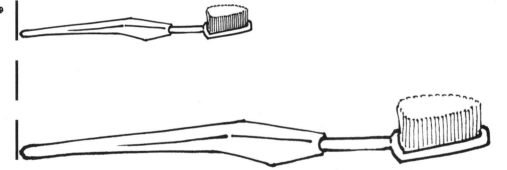

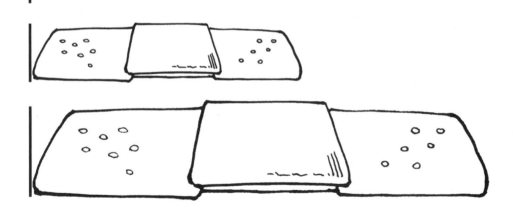

🐟 Draw a comb so that the objects are in order from shortest to longest.
🐢 Draw a toothbrush so that the objects are in order from shortest to longest.
★ Draw a bandage so that the objects are in order from shortest to longest.

CW62   Challenge

Name _____

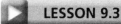

Up, Up, and Away

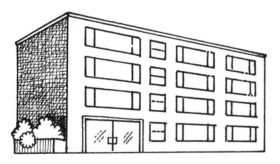

★

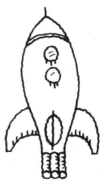

● Draw a tree that is shorter.
   🐢 Draw a building that is taller.
   ★ Draw a rocket that is shorter and a rocket that is taller.

Challenge   CW63

Name _____

LESSON 9.4

Which Object Is Tallest?

🐢

🐟 Find a crayon, a pencil, and a magic marker. Put them in order from shortest to tallest. Draw the objects from shortest to tallest and write 1, 2, and 3 to show the order. Circle the object that is tallest.

🐢 Find three blocks. Put them in order from shortest to tallest. Draw the objects from shortest to tallest and write 1, 2, and 3 to show the order. Circle the object that is tallest.

CW64 Challenge

LESSON 9.5

Name _____

Capacity in the Refrigerator

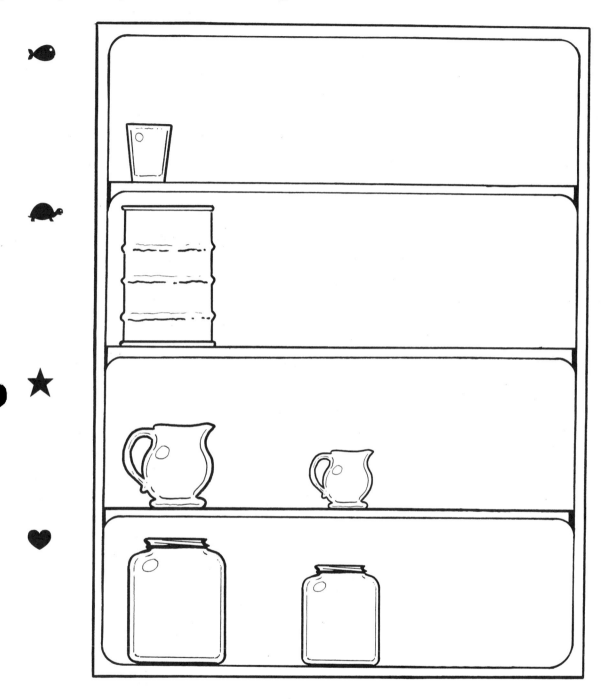

- 🐟 Draw a glass that holds more.
- 🐢 Draw a can that holds less.
- ★ Draw a pitcher that holds the most.
- ♥ Draw a jar that holds the least.

Challenge   CW65

Name _____

LESSON 9.6

Which Object Is Lightest?

 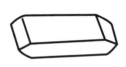 

- Hold each object. Mark an X on the object that feels heaviest. Underline the object that feels lightest.
- Draw the objects in order from lightest to heaviest.

CW66 Challenge

Name

LESSON 10.1

Shape Graph

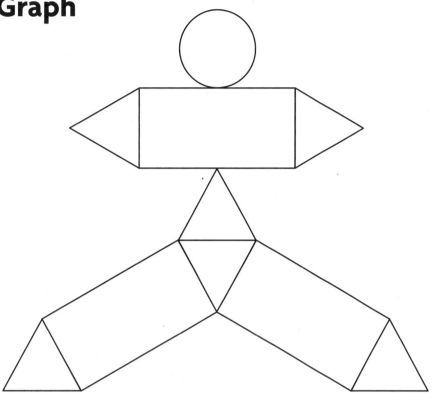

Plane Shape Person

Place plane shapes on the picture. Draw one of the shapes at the beginning of each row of the graph. Move the shapes from the picture into the correct rows of the graph. Draw the shapes on the graph.

Challenge CW67

Name _____

LESSON 10.2

Fruit Graph

Bananas and Coconuts

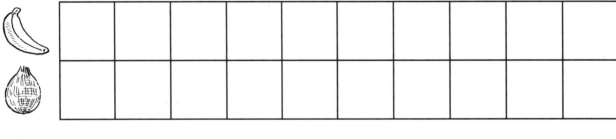

🐟 Look at the picture. What can you ask classmates about the picture? Use counters to make a graph. Draw the counters.

🐢 Write how many for each fruit.

CW68 Challenge

Name _____

Let's Play

How Children Play

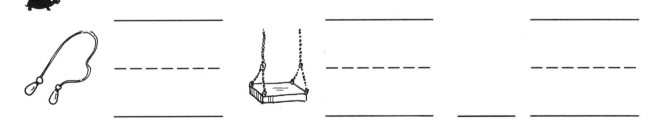

- 🐟 Draw a picture at the beginning of the bottom row of the graph to show one more way children are playing. Then use counters to make a graph.
- 🐢 Draw the third way of playing. Write how many children play in each way. Circle the number that shows the most. Underline the number that shows the fewest.

Challenge CW69

Name _____

> LESSON 10.4

Clothes Count

Classmates' Clothes

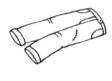

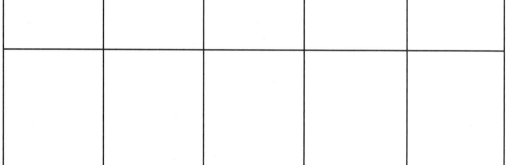

_____   _____   _____

- - - - -   - - - - -   - - - - -

_____   _____   _____

- What can you ask classmates about the clothes? Ask five classmates and draw a picture in the graph for each answer.
- Write how many of each kind.

CW70   Challenge

● Fruit Stand

Fruit We Like

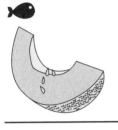

●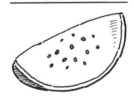

_____   _____   _____

- - - - - - -   - - - - - - -   - - - - - - -

_____   _____   _____

● 🐟 Look at the picture. What can you ask classmates about the fruit stand? Make a picture graph for the fruit.
🐢 Write how many of each kind. Circle the number that shows the most.

Challenge CW71

LESSON 10.6

Picnic in the Park

Do You Like Picnics?

Yes

No

———— Yes   ———— No

🐟 What can you ask classmates about picnics? Ask ten classmates if they like picnics. Make a concrete graph or a picture graph.

🐢 Write how many. Circle the number that shows more. Underline the number that shows less.

CW72   Challenge

Name _____

LESSON 11.1

Ways to Make 5

$0 + 5 = 5$ ___ + ___ = 5

___ + ___ = 5   ___ + ___ = 5

___ + ___ = 5   ___ + ___ = 5

Use two-color counters to model different ways to make 5. Write all the ways to make 5.

Challenge   CW73

Making 6 and 7

----- + ------ = 6

----- + ------ = 6

------ + ------ = 7

------- + ------ = 7

🐟 🐢 Use two-color counters to model different ways to make 6. Draw the counters and write the numbers.

★ ♥ Use two-color counters to model different ways to make 7. Draw the counters and write the numbers.

Name _____

LESSON 11.3

Making 8 and 9

🐟
```
_____ + _____ = 8
```

🐢
```
_____ + _____ = 8
```

★
```
_____ + _____ = 9
```

♥
```
_____ + _____ = 9
```

🐟 🐢 Use two-color counters to model different ways to make 8. Draw the counters and write the numbers.

★ ♥ Use two-color counters to model different ways to make 9. Draw the counters and write the numbers.

Challenge CW75

Name _____

Making 10

🐟
[   ]

_____   _____

`------` + `------` = 10

_____   _____

🐢
[   ]

_____   _____

`------` + `------` = 10

_____   _____

★
[   ]

_____   _____

`------` + `------` = 10

_____   _____

♥
[   ]

_____   _____

`------` + `------` = 10

_____   _____

🐟 🐢 ★ ♥ Use two-color counters to model different ways to make 10. Draw the counters and write the numbers.

CW76 Challenge

Name _____

LESSON 11.5

Circle How Many in All

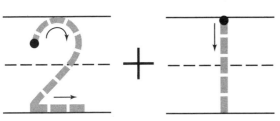

2   (3)   4

___ + ___

4   5   6

___ + ___

5   6   7

___ + ___

8   9   10

🐟 🐢 ★ ♥   Write how many in each group. Use counters to add. Circle how many in all.

Challenge   CW77

Name _____

LESSON 11.6

Add, Add, Add

3 + 3 = 6   (3 + 4 = 7)

6 + 1 = 7   7 + 1 = 8

 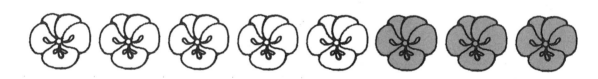

5 + 3 = 8   4 + 5 = 9

3 + 3 = 6   4 + 4 = 8

🐟 🐢 ★ ♥ Circle the addition sentence that matches the picture.

CW78 Challenge

Adding with Pictures

___ + ___ = ___

___ + ___ = ___

___ + ___ = ___

🐟 🐢 ★ Write how many in each group. Add. Write how many in all.

Challenge CW79

Trade a Story

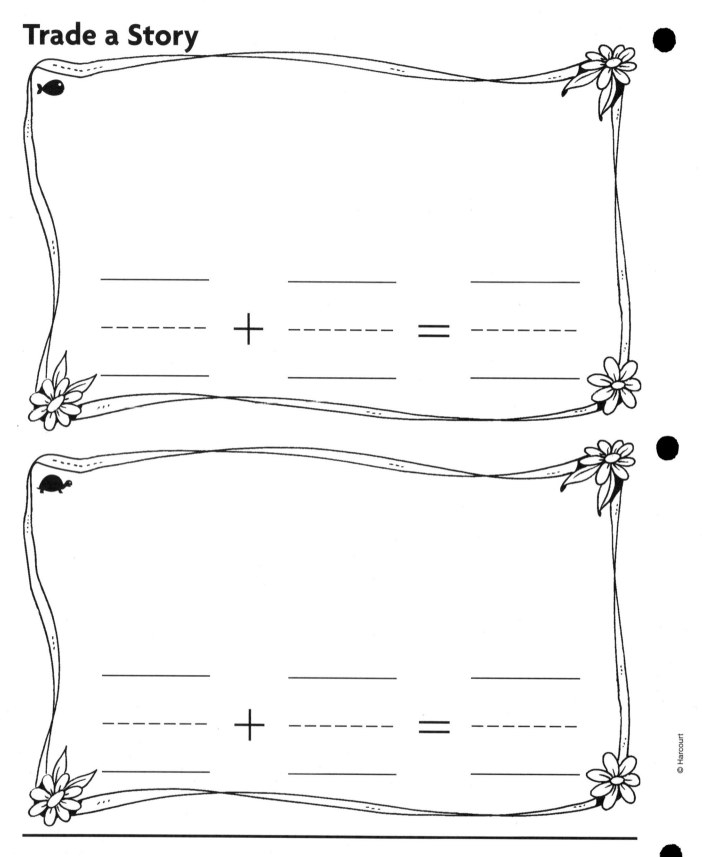

🐟 🐢 Draw a picture to show an addition story. Trade papers with a partner. Complete the addition sentence for your partner's picture.

LESSON 12.1

Name _____

● **How Many Are Left**

🐟

_____   take away   _____   _____

●

────────────────────────────

🐢

_____   take away   _____   _____

●

🐟 🐢 Tell a subtraction story. Model the story with connecting cubes. Write the numbers.

Challenge   CW81

How Many are Left?

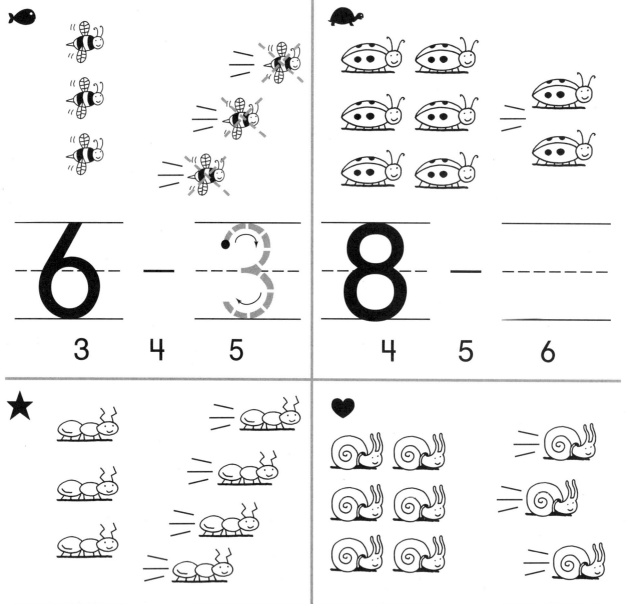

🐟🐢⭐♥ Use cubes to act out the picture. Mark an X on the ones that leave. Write how many leave. Circle the number that shows how many are left.

Name _____

 LESSON 12.3

Making a Subtraction Pattern

🐟 9 − 1 = 8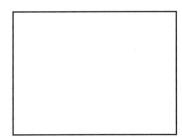

🐢 8 − 1 = _____

★   7 − 1 = _____

♥   6 − 1 = _____

✿   5 − 1 = _____

🐟 🐢 ★ ♥ ✿ Subtract. Make a picture to model the subtraction problem. Write the number that tells how many are left.

Challenge CW83

Dot Subtraction

🐟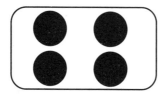

_____ - 1 = ☐

🐢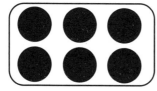

_____ - 2 = ☐

★

_____ - 3 = ☐

🐟 🐢 ★ Use counters to show how many there are in all. Write how many in all. Mark an X on the ones that are taken away. Write the number to show how many are left.

CW84 Challenge

Name _____

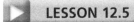

● **Subtraction Sentences**

_____   _____   _____

_____ take away _____ is _____

_____   _____   _____

_____   _____   _____

_____ take away _____ is _____

_____   _____   _____

★
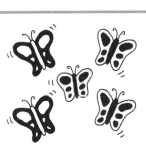   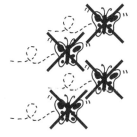

_____   _____   _____

_____ take away _____ is _____

● _____   _____   _____

 ★ Tell the subtraction story. Then complete the subtraction sentence.

Challenge   CW85

Name _____    LESSON 12.6

Drawing Pennies

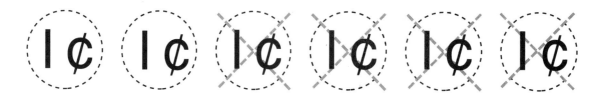

$6 - 4 = \underline{}$

$8 - 5 = \underline{}$

$9 - 3 = \underline{}$

🐟 🐢 ★ Draw the number of pennies in all. Mark an X on the number of pennies subtracted. Write how many pennies are left.

CW86 Challenge

Name _____

▶ LESSON 12.7

● Supermarket Subtraction

● 🐟   _____   _____   _____

 _____ take away _____ is _____

 _____   _____   _____

🐢   _____   _____   _____

 _____ take away _____ is _____

 _____   _____   _____

● 🐟 🐢 Look at the picture. Tell a subtraction story about some items in the supermarket. Draw a circle around each group of items that is in your story. Mark an X on the items you subtract. Then complete the subtraction sentence.

Challenge   CW87